11.90

5

SEE INSIDE

A SPACE STATION

Series Editor: **R. J. Unstead**

Warwick Press
New York/London/Toronto/Sydney
1988

CONTENTS

Author
Robin Kerrod

Published in 1988 by Warwick Press,
387 Park Avenue, New York, New York 10016.
Revised edition first published in 1988 by
Kingfisher Books Limited.

Copyright © Grisewood & Dempsey Limited 1977,
1988

Printed in Hong Kong

Library of Congress Catalog Card NO. 87-51052
ISBN 0-531-19031-5

The Challenge of Space

On October 4, 1957, a new moon appeared in the sky. It was a tiny, man-made moon launched into space by Soviety scientists, who called it "Sputnik." Sputnik heralded the start of an exciting new age for mankind—the Space Age. Less than four years later, on April 12, 1961, the Soviet cosmonaut Yuri Gagarin circled once around the earth in a spacecraft. A human being had taken the first tentative steps that would eventually lead people to land on the moon.

The Apollo manned missions to the moon between 1969 and 1972 proved that human beings could travel long distances in space. After Apollo, the Americans switched their attention to an orbiting space station called Skylab, which they launched in 1973. By then the USSR had already launched experimental space stations of their own, called Salyut. By 1984 they had launched seven Salyuts, each one an improvement on the one before. And Salyut cosmonauts had spent periods of up to $7\frac{1}{2}$ months in orbit, with no long-term ill-effects.

In 1986, the USSR launched a new-generation space station, Mir, which is the base unit for a large space station complex. The United States is planning an even more advanced station for the 1990s, which will include units provided by Europe and Japan. Like Mir, the NASA station will be manned continuously. In these stations scientists will keep an eye on the world's weather, try to discover hidden mineral deposits, and observe the heavens. They will also carry out all manner of experiments in the airless and weightless conditions of space. Later, the space stations will take on another role—as spaceports—to handle spaceships traveling to the moon and planets, and to space cities.

Right: Only a hundred years ago, space travel was an impossible dream, a matter for fiction and fantasy. This space ship was drawn by Gustave Doré in 1865 to illustrate Jules Verne's story, "From the Earth to the Moon."

Opposite: Clothed from head to toe in a bulky spacesuit, astronaut Jack Lousma ventures into the void of space. Reflected in the gold-tinted visor of his helmet is the bulk of the first big orbiting space station—Skylab.

APOLLO TELESCOPE MOUNT
This windmill-shaped unit carried four huge solar cell arrays, which provided much of Skylab's electrical power. At its center was the cluster of instruments for observing the Sun.

Solar panel

Apollo spacecraft

Skylab

Many years before people could actually fly in space, they had shown how it could be done—provided there were powerful enough rockets. The Soviet Konstantin Tsiolkovsky and the German Hermann Oberth were two of the greatest of the early space pioneers. The most far-sighted of the space enthusiasts saw that a vital step in the conquest of space would be to build a space station in orbit around the earth. But they had one fear. Could people survive long periods in space? If not, space stations and travel to other planets would be impossible.

By early 1973 the longest anyone had spent in space was 24 days. Then the United States launched Skylab, an experimental space station that would give astronauts the opportunity of staying in space for nearly three months—if they could stand it. Skylab was lifted into orbit by the powerful Saturn V Moon rocket. It circled the earth 270 miles (435 km) up, and could be clearly seen from earth as a bright moving star.

Three teams of astronauts visited Skylab in turn, shuttling back and forth in Apollo spacecraft from the Kennedy Space Center, in Florida. The picture here shows Skylab with an Apollo spacecraft linked to it. From nose to tail, the cluster was 119 ft. (36 m) long and weighed 90 tons.

The Skylab crews spent most time in the Orbital Workshop. Their living quarters were divided into sections.

One was for eating and relaxing off-duty; another was for sleeping. A third contained washing and toilet facilities, while a fourth contained experiments. Above the living quarters was a huge compartment containing experimental gear, spacesuits, storage lockers, and other equipment. The forward compartment was connected by an airlock to a docking unit. Astronauts used the airlock to seal off the living area from the rest of the space station. They could then open the airlock to space without all the air in the station escaping. The docking unit was fitted with two ports, or circular doors, in which spacecraft could dock, or link up together.

Skylab contained all of the features that future space stations must have. And the Skylab astronauts showed that people could survive happily for periods of up to 84 days in space.

Above: This Apollo spacecraft was used to ferry the Skylab astronauts to and from the space station. The command, or crew module at the front carried the three-man astronaut crew.

Right: One of the Skylab experiments was to see if the spider "Arabella" could spin webs in weightless conditions. She did quite well, as you can see.

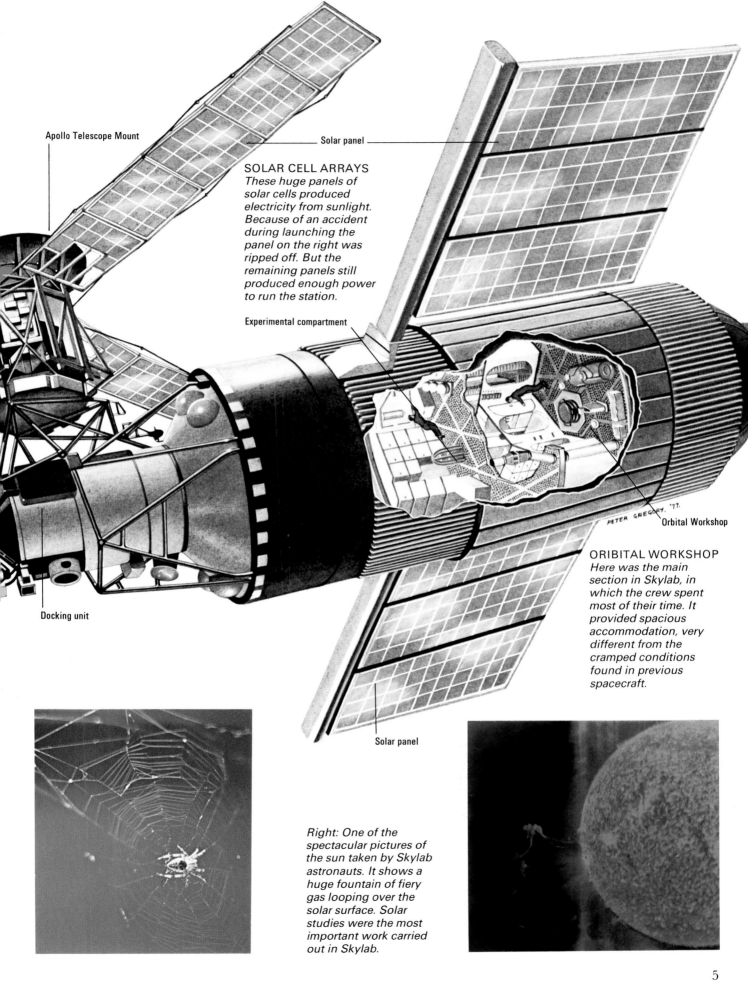

Apollo Telescope Mount

Solar panel

SOLAR CELL ARRAYS
These huge panels of solar cells produced electricity from sunlight. Because of an accident during launching the panel on the right was ripped off. But the remaining panels still produced enough power to run the station.

Experimental compartment

Orbital Workshop

PETER GREGORY. '77.

ORIBITAL WORKSHOP
Here was the main section in Skylab, in which the crew spent most of their time. It provided spacious accommodation, very different from the cramped conditions found in previous spacecraft.

Docking unit

Solar panel

Right: One of the spectacular pictures of the sun taken by Skylab astronauts. It shows a huge fountain of fiery gas looping over the solar surface. Solar studies were the most important work carried out in Skylab.

A Russian Salyut

Above: Little by little cosmonauts in the Soyuz craft on the left are edging toward a Salyut space station (right). Soon they will be firmly docked, or linked together. The cosmonauts will transfer to Salyut.

Below: A Soyuz spacecraft leaving the launch pad, bound for a rendezvous in space with a Salyut space station.

The USSR launched the first of a series of Salyut space stations in 1971. The most successful of them all has been Salyut 7, launched in 1982. In this craft cosmonauts Leonid Kizim, Vladimir Solovyov and Oleg Atkov spent a record 237 days in orbit, beating all previous space duration records. Salyut 7 was still in orbit and habitable in 1986, when the new space station Mir went into orbit (see page 8).

Salyut 7 is much smaller than Skylab was. It is about 46 ft. (14 m) long, and has a diameter of 13 ft. (4 m) at its widest point. Two or three cosmonauts usually man the craft, although it sometimes has extra "visiting" cosmonauts from countries other than the USSR. Large panels of solar cells provide it with electricity from sunlight.

The cosmonauts carry out their experiments and observations in the large working compartment. In the wide part of the craft are two wall beds and two lavatories. The middle section houses a pantry, a table, and seats. In the front is the main communication and control panel with seats for both pilot and copilot. Above the control panel is a hatch. This hatch leads to a compartment that links with the spacecraft that ferries crews to and from the space station.

This spacecraft, called Soyuz, is about 23 ft. (7 m) long. It is made up of three main sections, or modules. The crew travel to and from space in the spherical descent module in the front. This has a docking hatch that links with a hatch on Salyut. Connected to the descent module is a larger orbital module. This is where the crew live and work while they are in orbit.

Salyut 7 has one docking hatch in front and one at the rear, so that two spacecraft can link up with it at the same time – often a Soyuz and a supply craft called Progress, which is unmanned.

Returning to Earth

The descent module is the only part of Soyuz that returns to earth. With the cosmonauts strapped inside, it re-enters the earth's atmosphere at speeds of over 17,000 mph (27,000 kilometers an hour). The atmosphere slows it down, then *retro-rockets* fire beneath it to slow it down further. Finally, a parachute opens to lower it gently to the ground. Unlike the American Apollo astronauts who splashed down at sea, Soviet cosmonauts always come down on land.

Both Soyuz and Salyut spacecraft are launched from the main Soviet space center, the Baikonur Cosmodrome. It is located near the town of Leninsk, in Kazakhstan, more than 1,300 miles (2,000 km) southeast of Moscow.

The interior of a Salyut craft, showing one of the crew at the control console.

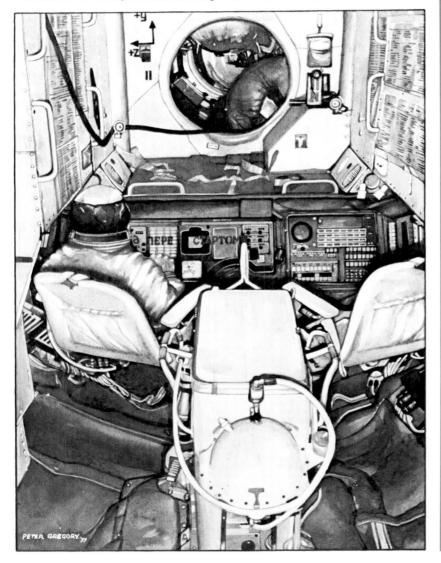

Astronaut Deke Slayton poses with the Soyuz commander Alexei Leonov during the joint Apollo-Soyuz flight. Leonov made the first ever space "walk" in 1965.

The Space Handshake

On July 17, 1975 the space rivalry between the United States and Russia was forgotten as American astronauts and Soviet cosmonauts shook hands in orbit 137 miles (220 km) above earth. The occasion was the first international link-up between an American Apollo and a Soviet Soyuz spacecraft.

In orbit the two craft were linked by means of a special docking unit, each end of which fitted one of the craft. Units like this are needed before one kind of craft can link up with another, because each has a different kind of docking system. Future space stations may be fitted with a standard assembly that every craft will have. The use of the standard unit will make space rescue missions a lot easier.

The Apollo-Soyuz joint flight was a great success even though the two craft were directed separately from mission control centers in Texas and Moscow in two languages.

Space Station Mir

The USSR's success with their Salyut space stations, particularly Salyuts 6 and 7, led them to launch a new type of station in early 1986. It is called Mir, meaning Peace. From the outside, Mir looks rather like Salyut 7, and it is much the same size, but inside it is very different.

The inside of Salyut is very cramped indeed. It is filled with equipment with which the cosmonauts carry out their experimental work. There is little room for crew living quarters. Mir, on the other hand, is fitted out mainly as living quarters. It also contains the necessary communications and control consoles, but there is little in the way of experimental equipment. The living quarters of Mir are divided into separate small cabins for each crew member, male and female. They each contain a chair, a desk, a sleeping bag, and storage space.

Mir also differs from Salyut in another way. It has a spherical docking module at one end, which has five ports (hatches) to which spacecraft can link up. There is another docking port at the other end. This means that as many as six spacecraft can link up with Mir in orbit. In this way, Mir can be the base unit for a large space station complex.

Two or more of the docking ports will be kept free so that ferry craft such as Soyuz and Progress can dock with Mir. Scientific and laboratory modules can dock at the other ports. Cosmonaut-scientists work in these modules, but have their living quarters in Mir.

The first science module to link up with Mir docked with it in April 1987. It is called Kvant, named after the physics term "quantum." It is a cylinder about 20 ft. (6 m) long and 13 ft. (4 m) across. Kvant was launched into orbit with a booster unit attached. It was unmanned, and its flight was controlled from the Soviet mission control. When it had docked with Mir, the booster separated. This left another docking port free to receive another spacecraft.

The first cosmonauts to occupy Mir were Leonid Kizim and Vladimir

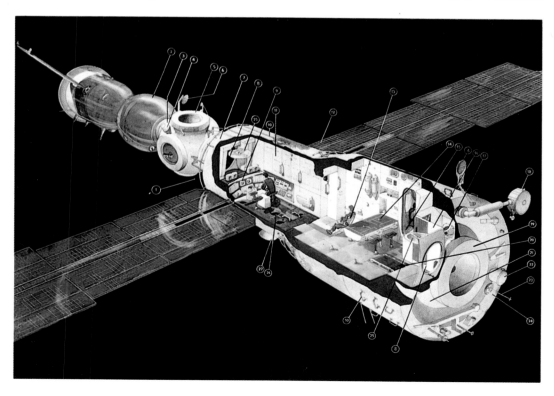

A cutaway of the Mir space station, showing it docked with a Soyuz ferry. A crew of three are shown aboard, but it can accommodate up to six. This is an original Russian illustration, and not all the technical information is available, but among the most interesting items numbered are the following:

2. Soyuz ferry ship
5. Docking module
8. Transfer hatch
9. Control console
10. Hand rails for EVA
11. Storage lockers
12. Solar panels
14. Chart table
15. Crew cabin
19. Rear docking port

Many of the other items numbered are antennas or docking targets.

THE SOVIET SHUTTLE

Even now, the USSR still launches its cosmonauts into space in Soyuz spacecraft, the first of which flew in 1967. But in future years, cosmonauts will travel into orbit by shuttle. The Soviet shuttle is often called Kosmolyot.

In fact, there appear to be two Soviet shuttles, one large and one small. However, it is possible that the small one is a test vehicle for the larger one. The small one has been flight tested in space several times, and has been photographed being recovered. Photographs have also been taken of the larger version, which looks much like the American shuttle (see page 10).

Unlike the American shuttle, the Soviet shuttle does not have its main rocket engines in the tail. The Soviet shuttle is thrust into orbit by the engines of a booster rocket. The booster will probably be a rocket known as Energia, which made its first flight in the summer of 1987. It is the most powerful rocket ever launched. Measuring about 220 ft. (67 m) long, it can place in orbit a payload (cargo) of more than 98.4 tons (100 tonnes). This is three to four times the payload that the American shuttle can launch. As well as launching the shuttle, Energia has the power to carry into orbit bigger modules for space stations and sections of large spacecraft that will fly on interplanetary missions.

Solovyov. After 75 days in Mir, they rocketed over to Salyut 7, where they spent 50 days before returning to Mir. By the time the cosmonauts returned to earth in July, Kizim had broken another record. He had become the first person ever to spend a total of more than a year in space.

It is likely that a number of Mir space stations will eventually be launched, with each one being used for a different purpose. One will be a scientific station, made up of modules like Kvant. Another may be a satellite service station equipped to launch and repair satellites. There could also be an interplanetary station, which would build large spacecraft for traveling to the moon and the planets from sections launched separately into orbit.

Mars would be the first target for a planetary spaceship, and the USSR is certainly planning one for launch early on in the next century. Maybe the United States will join them in sharing the costs and solving the many technical problems such a long trip will present.

Russia's mighty rocket Energia blasts off the launch pad on 15 May 1987 on its impressive maiden flight.

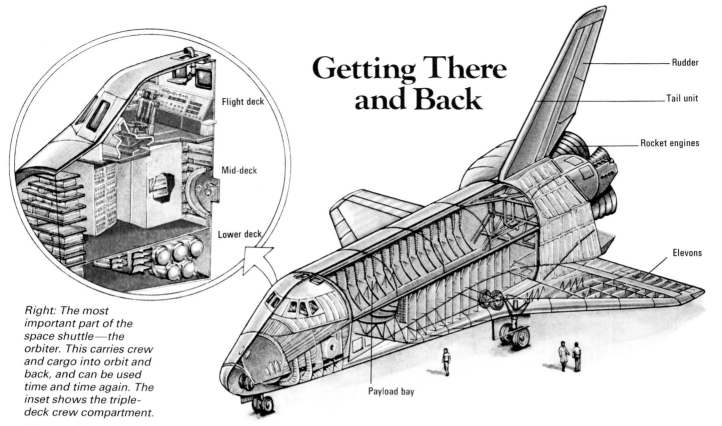

Getting There and Back

Rudder

Tail unit

Rocket engines

Elevons

Flight deck

Mid-deck

Lower deck

Payload bay

Right: The most important part of the space shuttle—the orbiter. This carries crew and cargo into orbit and back, and can be used time and time again. The inset shows the triple-deck crew compartment.

The main means of ferrying up crews and supplies to space stations in the future will be by shuttle craft, which are able to make repeated journeys. The American space shuttle was the first re-usable craft to fly, in 1981.

The main part of the space shuttle is the orbiter, which is about the size of a medium-range airliner. It takes off like a rocket but lands like an aircraft. It normally carries a crew of up to seven astronauts, both men and women. There are at present three orbiters in the shuttle fleet, named Columbia, Discovery and Atlantis. A fourth orbiter, named Challenger, tragically exploded during its launch in January 1986, killing the seven astronauts on board. Another orbiter is being built to replace it.

On the launch pad, the orbiter sits on its tail, mounted on top of a huge tank that carries fuel for its three main rocket engines. Strapped to the tank are two booster rockets. All the rockets fire to launch the shuttle, but two minutes after lift-off the boosters separate and parachute back to earth. They are recovered, and can be used again. A few minutes later, the main engines cut out as the fuel tank runs dry. It separates, but is not recovered.

The orbiter continues into orbit, about 150 miles (250 km) high. There it can do a variety of things, depending on its payload, or cargo. Usually it carries satellites into orbit, but sometimes it has a special payload such as Spacelab (see page 14). At other times it is used on exciting satellite rescue missions. The payload bay is the biggest part of the shuttle, measuring more than 60 ft. (18 m) long and 15 ft. (4.5 m) in diameter.

Right: Shuttle orbiter Atlantis blasts off the launch pad on its maiden flight in 1985.

Below: From lift-off to landing. Two minutes into the flight, the boosters separate (1). The orbiter's main engines continue firing until the fuel tank runs dry and falls away (2). Two small engines then fire to boost the craft into orbit at 17,500 mph (28,000 km an hour). To descend from orbit, the orbiter fires its retro-rockets. It reenters the atmosphere and slows down (3). Finally it lowers its landing gear and touches down on a runway (4).

1

2

3

4

Turnabout

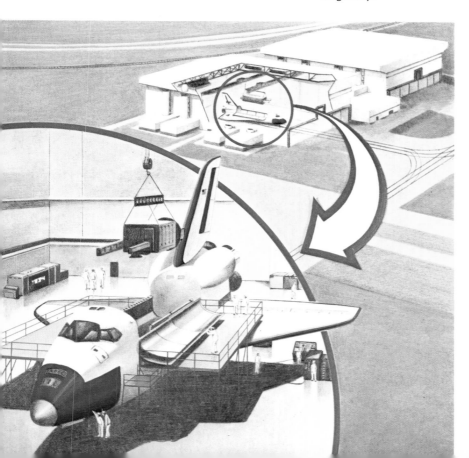

Above: An orbiter about to land after its stay in space. As soon as it has touched down it is made safe, and any experimental equipment is removed.

Below: The orbiter is being serviced in the Orbiter Processing Facility, and a new experimental unit is being lifted into the cargo bay.

When the shuttle orbiter has completed its mission, it reenters the earth's atmosphere and glides to a landing on a runway like a normal plane. But it lands much faster, so the runway is much longer than is usual at an airport. At the shuttle's home base, the Kennedy Space Center, the runway is over 3 miles (5 km) long.

As soon as the orbiter has landed, it undergoes *safing* in a huge hangar called the Orbiter Processing Facility. "Safing" means removing the explosive rocket fuels that remain in the fuel tanks and pipes. The orbiter is also serviced for the next mission.

All mechanical, electrical, and hydraulic (liquid pressure) systems in the craft are checked and replaced if necessary. The engines are examined for wear. So is the outer surface of the orbiter. This surface acts as a heat shield when the returning craft plunges through the atmosphere at high speed. It protects the orbiter from the heat caused by friction with the air. It is made of tens of thousands of heat-resistant tiles, any of which can easily be replaced.

The orbiter is then towed to the huge Vehicle Assembly Building. Two of

building's four high bays are used to assemble the shuttle system for launching. The other two bays are used to make ready the twin rocket boosters and fuel tank on which the orbiter rides into space.

When the orbiter, boosters and tank are ready, they are lifted into their launch position on a mobile launch pad. The huge doors of the Vehicle Assembly Building roll back, and the biggest vehicle in the world emerges carrying the launch pad. The vehicle measures 131 ft. by 114 ft. (40 m by 35 m), and has twin caterpillar tracks at each corner. It carries the pad at a snail's pace to the launching site. At the site is a tall tower, from which engineers can reach the various parts of the shuttle. Another unit swings into position over the shuttle to enable the crew to enter and new payloads to be installed. It swings back just before the shuttle blasts off.

Below: Here the orbiter is seen leaving the Vehicle Assembly Building after it has been mounted in its launching position on top of the fuel tank.

Laboratory in Space

One of the space shuttle's most important payloads is Spacelab. This is a fully-equipped laboratory that fits into the shuttle's payload bay, where it stays throughout the mission. Like the shuttle, it is designed for repeated use. The first Spacelab mission took place in November 1983.

Spacelab can be varied from flight to flight by joining together different units, or modules. The main unit is the laboratory module, which contains the main scientific instruments and experimental equipment. This section is pressurized, like the shuttle's crew quarters, so scientsits are able to work there in ordinary clothes instead of space suits. The other part of Spacelab consists of pallets, which are open to space. The pallets carry telescopes and other instruments.

EUROPE'S "NASA"
Spacelab is designed and built by ESA, the European Space Agency. This agency coordinates space activities in 13 European countries—Austria, Belgium, Britain, Denmark, France, West Germany, Ireland, Italy, Norway, the Netherlands, Spain, Sweden, and Switzerland. ESA has its headquarters in Paris and major centers at Nourdwijk (Netherlands) and Darmstadt (West Germany).

Left: On the Spacelab D-1 mission in 1985, a West German scientist bares his arm ready for a blood sample to be taken. By regularly checking the state of the blood in orbit, doctors can find out how space affects the human body.

Below: A scene inside the laboratory module during the first Spacelab mission in 1983. On the right is Dr. Ulf Merbold, who became the first European to fly on an American spacecraft.

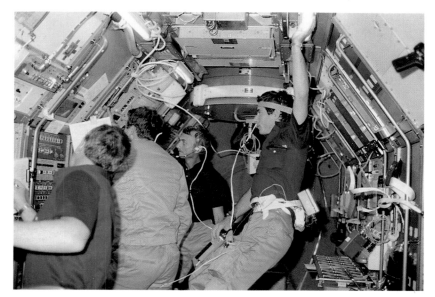

The usual laboratory module flown on Spacelab missions consists of a cylinder 23 ft. (7 m) long and 13 ft. (4 m) in diameter. It is connected by a tunnel with the crew compartment on the mid-deck of the shuttle orbiter, where the Spacelab scientists live when they are not working inside the laboratory. They are highly qualified men and women, specially trained to carry out the experiments being performed on the mission.

The experiments are carried out in the life sciences (biology and medicine), material sciences, astronomy and earth observations. Doctor-astronauts use the crew as guinea pigs to see how space flight affects the human body. They take blood samples and carry out many other tests. Other scientists experiment with making new materials. They find that, in the weightlessness of space, they can make super-strong alloys.

Below: A cutaway picture which shows what Spacelab looks like when it is up in orbit. Here it is made up of the laboratory module and two pallets carrying instruments that need to be exposed to space. Note that one of the astronauts is space-walking to check the instruments.

Living in Orbit

When people venture into space, they must take an exact copy of earth's atmosphere with them. They must also take food to eat and water to drink. They must have somewhere to sleep, wash and go to the lavatory. They must have medical supplies in case they become ill. All these things are provided by a spacecraft's life-support system.

Spacecraft are usually provided with an atmosphere of oxygen and nitrogen much like that on earth, though often at a lower pressure. An air conditioner removes smells, moisture, and carbon dioxide from the air. It also keeps the temperature of the air at a comfortable level. Drinking water may be carried into orbit in tanks or it may come from fuel cells. Fuel cells produce electricity

by making hydrogen and oxygen gases combine to form water.

The condition of weightlessness, or no gravity, is the most difficult one for astronauts to get used to. While it may be fun to be weightless and be able to do endless cartwheels, it can soon become tiresome. Also, constantly being weightless can make the body muscles weak. Astronauts on long space missions must therefore take regular exercise—for example on cycling devices.

In future, large space stations will be rotated to create an artificial gravity. Things will be flung outward toward the inner walls by centrifugal force. This will imitate gravity and give the crew a feeling of weight and of "up" and "down."

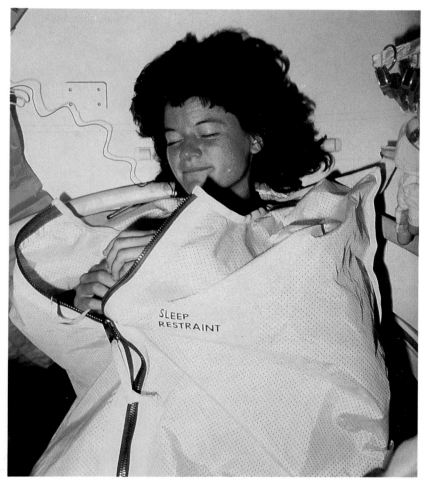

But for the foreseeable future, astronauts have to change the way they do things when they venture into orbit. They spend a great deal of time just floating in the air. If they want to stay in the same place for long, they must anchor themselves down with suction shoes or foot straps. When they sleep, they must do so inside sleeping bags firmly anchored to bunks or the sides of the spacecraft.

Personal hygiene also presents quite a problem in zero gravity. Ordinary washing would send showers of water droplets everywhere, so astronauts have to be content with a rub down with a wet towel. Going to the lavatory is another problem when there is no gravity to carry away the body wastes. But space engineers have designed a lavatory for the shuttle that solves the problem. It is "flushed" by a stream of air, which removes the wastes.

Above: The first American woman astronaut Sally Ride, ready for bed. Like other astronauts, she sleeps zipped inside a sleeping bag, which is anchored to part of the spacecraft to prevent it floating away.

Above left: To keep fit while in space, the astronauts take regular exercise on treadmill devices. They wear an elastic harness to hold them down to the floor as they run on the spot.

Eating and Drinking

The early astronauts had to eat the most unappetizing food, squeezed into their mouths from toothpaste-type tubes. However, things are different today. On the shuttle, for example, more than seventy different foods are available. The astronauts can breakfast on scrambled egg and sausage, lunch on ham and pineapple, and dine at night on soup, turkey, and strawberries. And they eat their meals with an ordinary knife, fork, and spoon.

The foods may come in natural forms, or be canned or dehydrated—that is, with the water removed. The dehydrated foods have to be mixed with water before serving. Drinks must be sucked through a tube or squirted into the mouth.

Space Walk

Astronauts spend most of their time in orbit inside their spacecraft, where there is an earth-type atmosphere. Occasionally they may have to venture outside their craft into space. We often call this "taking a space walk," but the proper name for it is Extra-Vehicular Activity, or EVA. Astronauts perform EVAs to test new equipment, to change films in automatic cameras, and to carry out running repairs on their spacecraft. Soon astronauts will also be involved in constructing space stations and other large structures in space.

When performing EVA, astronauts take their atmosphere with them in a spacesuit. The suit is made up of several different layers. Next to the skin is a garment through which water is circulated. It keeps the astronaut cool. On top of this is a pressure garment, which supplies oxygen under pressure. It has a transparent helmet, over which fits a visor. This shields the astronaut from strong sunlight. The pressure garment is made up of several layers of heat-insulating material, which also afford protection against tiny meteoroid particles and harmful rays.

The modern spacesuit, like the one used on the shuttle, is self-contained. The oxygen for breathing and the water for suit-cooling are supplied by a life-support backpack built into the upper part of the suit. Astronauts control their suits from a chest unit.

To move about in space, astronauts wear a jet-propelled backpack, which on the shuttle is called a manned maneuvering unit (MMU). This unit which fits over the astronaut's spacesuit, is powered by tiny jets of pressurized gas. It is controlled by handles at the end of its armrests. Moving the handles fires different sets of jets.

Opposite: Today's astronauts travel freely in space, using jet-propelled backpacks like the MMU. This is Bruce McCandless.

Right: Edward White made the first American space walk in 1965, tethered to his spacecraft.

Below: Using the MMU, astronauts can fly to satellites already in orbit and capture them. Then they can jet them back to a spacecraft for repair.

Mission Control

Traveling to and from space has become quite a regular routine. But it still remains a very dangerous activity. People are shot into space in a flimsy metal "container" on top of thousands of tons explosive fuel. Their container, or spacecraft, is a complicated machine made up of thousands upon thousands of parts. Some of these can and do go wrong from time to time.

To reduce the dangers of space flight, nothing is left to chance. Vital systems in the spacecraft and rocket are duplicated, so there is less chance of them failing. Flight operations are worked out in detail and checked and re-checked. Astronauts train for their mission in *simulators*, or dummy spacecraft, that perform on earth like the real spacecraft do in space. When the astronaut's real mission is in progress, it is essential to make sure that their spacecraft is working properly. This

is the main task of the *mission control* team on the ground.

At the mission control center teams of controllers keep a close watch on how a flight is progressing. They sit at operating consoles equipped with TV-type monitoring screens, and other display devices, tape recorders, and communications links.

Each flight controller has a certain responsibility. One is concerned with the flight path of the spacecraft. Another checks that the spacecraft's mechanical and electrical systems are working well. A medical controller watches over data that tell him how fit the crew are. Others are concerned with navigation and guidance; with space walks; with experiments; and so on. In overall charge of mission control is the *flight director*. He instructs the astronauts through CapCom.

KEEPING TRACK

By means of a worldwide network of tracking stations, mission controllers can keep in contact with spacecraft anywhere in space. The American tracking network has stations in North and South America, Spain, Australia, and on the islands of Hawaii, Madagascar, Guam, and Ascension Island. Russia has ground stations scattered over its vast sprawling land mass. It also has a number of tracking ships to extend its coverage all around the world.

Left: 1 *CapCom (capsule communicator) communicates with the astronauts. He passes on questions and instructions from the flight director, the man in charge of the whole operation.*
2 *The medical controller observes the stresses imposed on the astronauts by the flight.*
3 *The flight dynamics controller checks the flight data and flight path of the spacecraft.*
4 *The systems controller keeps a close watch on the various systems that make up the spacecraft and ensures that they are all working properly.*

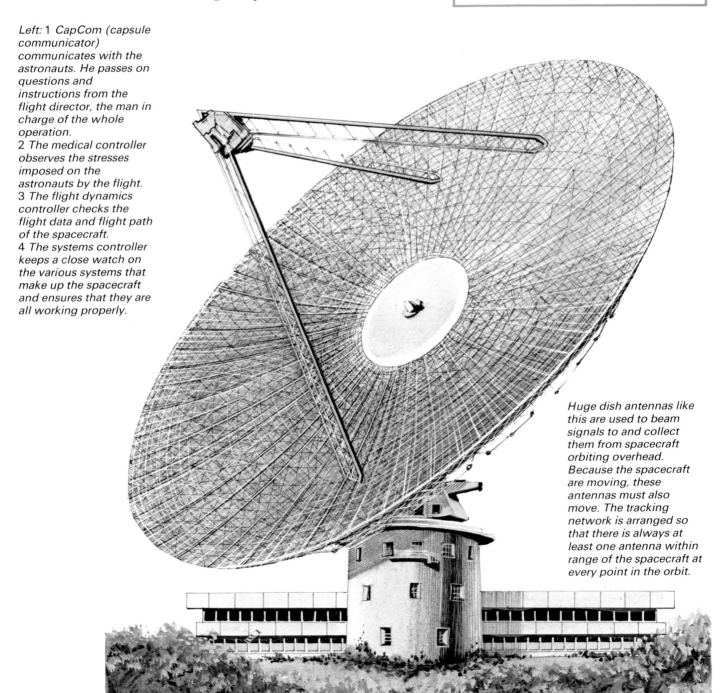

Huge dish antennas like this are used to beam signals to and collect them from spacecraft orbiting overhead. Because the spacecraft are moving, these antennas must also move. The tracking network is arranged so that there is always at least one antenna within range of the spacecraft at every point in the orbit.

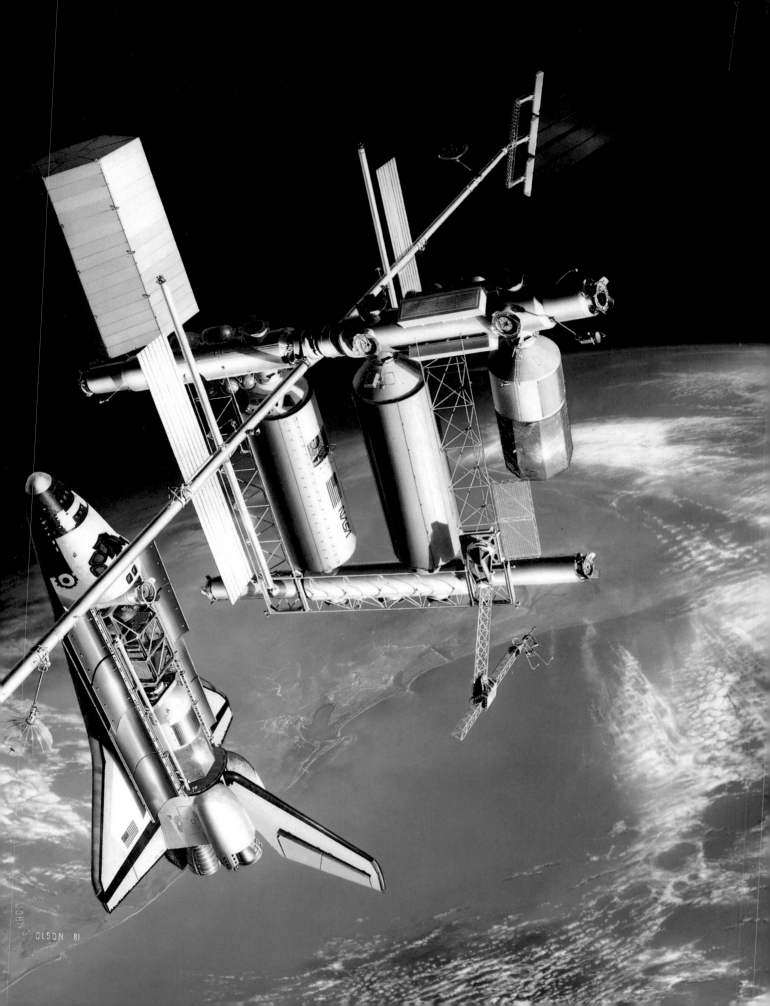

The NASA Space Station

By the end of this century, the Mir space station (see page 8) will have grown into a large complex, composed of six or more modules. Another large space station will also be orbiting the earth by then. It will be NASA's space station, which should be started in the mid-1990s.

NASA will do most of the construction work and provide most of the units, or modules, of which it will be made. However, Europe, Japan, and other countries will also take part in the station in one way or another. Europe, for example, will provide one of the modules and other equipment in a project named Columbus. It will use its heavy rocket launcher Ariane to launch some of the units.

The NASA station will be built up on a "keel" of one or two aluminum girder frames. The other sections will be built out from the keel. They will include modules for living, working, and for storing supplies. The modules will be about 60 ft. (18 m) long and 13 ft. (4 m) across. They will be designed to fit into the bay of the shuttle, which will ferry them up into space.

In orbit, the modules will be connected to the keel and to one another by astronaut-engineers. These skilled spaceworkers will fly about in MMUs (manned maneuvering units) and use keel-mounted cranes for the assembly work. Another major station unit will provide power. The main parts of this will be two huge panels of solar cells, each the size of a football field, which will convert the sunlight into electricity.

When the station becomes fully operational, up to eight astronauts will spend periods of up to three months working in the station. The space shuttle will visit the station at regular

Opposite: By the end of this century, the NASA space station could look something like this. It has two main modules, one for living in, the other for working. The unit on the right is a supplies module. The shuttle here is bringing up a module to replace it, containing fresh supplies.

Below: The NASA space station will eventually become a base for more adventurous space construction work. The picture shows a solar power satellite nearly finished. Note how big it is compared with the shuttle orbiter.

intervals to take them fresh supplies. The astronaut-scientists will carry out all kinds of experiments and observations up in orbit. Astronaut-service engineers will spend much of their time recovering and servicing satellites. They will travel to some satellites in "runabouts" called orbital maneuvering vehicles (OMVs). And they will send unmanned vehicles to travel to and bring back satellites from distant orbits they cannot reach themselves.

In the next century the station could be expanded into a spaceport. It would act as a base for building and servicing interplanetary spaceships, which would travel to the moon and Mars. On both these worlds there would by then be scientific bases and perhaps even mining colonies.

Space Cities

The building of permanent manned space stations will mark the true beginning of the conquest of space by people. But they will be much more than just a base for carrying out experiments and observing the heavens. They will act as a springboard for wider exploration of the universe.

As a base for making observations, the moon is ideal. It has no wet, dusty, atmosphere to dim and distort the feeble light from the stars. To start with, astronauts will build a base on the moon from used rocket bodies and spacecraft assembled in earth orbit But later they will be able to mine ores on the moon and smelt their own metals to provide materials for expanding the base.

The moon mines would also provide the raw materials for huge space cities which could be built early next century. These cities would be sited in a region of space about 240,000 miles (350,000 km) from earth. The illustration shows part of one space city design by the American scientist, Professor Gerard O'Neill. The design is big enough to house hundreds of thousands of people.

The main part of the space city, or colony, is two huge cylinders each about 20 miles (30 km) long. Three windows run lengthways down each cylinder, and sunlight is reflected into them by mirrors. Opposite the windows on the inside of each cylinder are three land areas where there is an earth-type landscape, with houses, trees, fields, and rivers. The cylinders are rotated to make an artificial gravity so that colonists can walk normally and the landscape stays in place. Inside the rim of the wheel are areas devoted to agriculture, laboratories, and more living accommodation. The climate is carefully controlled so that plants flourish and the people are healthy.

1 Wheel rim contains agricultural areas, laboratories, and living quarters
2 Fixed mirror reflector
3 Hub center (zero gravity zone)
4 "Outside" of mirror reflecting light onto landscapes B
5 Landscapes A
6 "Outside" of landscape area C
7 "Upside down" landscape B
8 "Outside" of mirror reflecting light onto landscape A
9 Radiator releases waste heat into space. All heat, light, and energy is derived from sunlight

IMPORTANT HAPPENINGS

	Unmanned Flight	Manned Flight
Early days	**1903** Konstantin Tsiolkovsky lays down the principles on which space flight depends. **1926** Robert Hutchings Goddard fires the first liquid-propellant rocket in the USA. **1941** A German team launches the A4 (V2) rocket, the ancestor of all modern rockets. **1949** The U.S.A. launches the first two-stage rocket, the V2/WAC Corporal, from White Sands rocket range. **1957** The USSR launches Sputnik 1 on 4 October. This is the world's first artificial satellite. **1958** On January 31, the U.S.A. launches its first satellite, Explorer. **1959** The USSR launches Lunik 1, the first successful probe to the moon.	**1955** The USSR begins construction of the Baikonur Cosmodrome. **1957** The USSR launches the first living thing into orbit. This is the dog Laika, in Sputnik 2. **1958** The U.S.A. announces its first manned space project, Mercury. In October, NASA is set up. **1959** NASA names the first seven astronauts for the Mercury project.
The 1960s	**1960** The U.S.A. launches the first communications satellite. This is a huge balloon called Echo 1. **1962** The U.S.A. launches Telstar, the first active communications satellite. They also launch the Mariner 2 probe to Venus, which reports on the furnace-like conditions there. **1964** The U.S. lunar probe Ranger 7 sends back thousands of close-up pictures of the Moon before crash-landing. **1965** The U.S. Mariner 4 space probe sends back the first close-up pictures of Mars, from a distance of 137 million miles (220 million km) in a fantastic feat of telecommunications. **1966** Luna 9 of the USSR makes the first instrument landing on the moon and sends back photographs. In May, the U.S. Surveyor probe makes a true soft landing, transmitting back thousands of high quality pictures. The U.S. probe Lunar Orbiter goes into orbit in August, and begins to map the whole lunar surface from orbit. **1967** The first flight of the Saturn V moon rocket takes place in November. **1969** NASA tests a nuclear rocket engine – the kind that may eventually be used for interplanetary flights.	**1961** Yuri Gagarin becomes the first man in space on April 12. He makes one orbit of the earth in his capsule, Vostok 1. In May, U.S. astronaut Alan Shepard makes the first sub-orbital flight. **1962** John Glenn becomes the first American in orbit, making three orbits of the Earth on February 20 in his capsule Friendship 7. **1963** The USSR launches the first woman in space, Valentina Tereshkova. **1964** The USSR launches Voshkod 1, a new spacecraft large enough to carry three cosmonauts. **1965** Soviet cosmonaut Alexei Leonov pioneers the art of spacewalking, from his spacecraft Voshkod 2 in March. Also in March, the U.S. two-man Gemini mission goes into orbit. In June, astronaut Edward White makes the first U.S. spacewalk from Gemini 4. **1968** The first flight of the Apollo spacecraft takes place, when Apollo 7 orbits the earth in October. In December, Apollo 8 makes a circumnavigation of the moon, paving the way for the moon landings. **1969** Apollo 10 makes a complete dress rehearsal for a lunar landing, while orbiting the moon in May. In July, Apollo 11 flies to the moon. Neil Armstrong and Edwin Aldrin touch down on 20 July, leaving Michael Collins in the Apollo mother ship. They spend $2\frac{1}{2}$ hours exploring the surface on foot.
The 1970s	**1970** The USSR launches a probe to Venus called Venera 7, which lands on the planet's surface. In November, the Soviet probe Lunokhod 1 becomes the first mechanical vehicle to travel over the moon's surface. **1972** NASA launches the first earth resources satellites, ERTS-1. This is later re-named Landsat 1. **1973** NASA probe Pioneer 10 sends back the first close-up pictures of Jupiter.	**1970** In April, the crew of Apollo 13 have a lucky escape when their spacecraft is crippled on the way to the moon. **1971** The USSR launches the first space station, Salyut 7, in April. **1973** In May, NASA launches its first space station, Skylab, which is almost crippled during its launch. However, the frst crew are able to repair it and make it habitable during their 28-day stay. The second Skylab crew occupy the space station in July, remaining for 59 days.

Unmanned Flight	Manned Flight
1973 The U.S. probe Mariner 10 becomes the first spacecraft to visit two planets—Venus and Mercury.	**1974** The Skylab 3 crew return to earth in February after a 84-day stay in orbit, during which they set a new space duration record. **1975** The U.S.A. and the USSR make the first joint space flight in the Apollo-Soyuz Test Project.
1976 U.S. Viking space probes go into orbit and drop landers to photograph the Martian surface and analyze the soil. Viking 1 lands in July.	**1977** The first free flight of the prototype space shuttle orbiter Enterprise takes place in August. It glides to a perfect landing after being released from its Boeing 747 carrier aircraft. In September, the USSR's most successful space station to date, Salyut 6, enters orbit and provides the home for long-stay cosmonauts. **1978** In January, Salyut 6 is linked in space with two Soyuz ferry craft, becoming the first triple spacecraft complex and pointing the way ahead. Later in the month, the first docking takes place between Salyut 7 and Progress, an automatic supply vessel.
1979 The U.S. probe Voyager 1 sends back fantastic close-up pictures of Jupiter and its moons in March. In September, Pioneer 11 sends back images of Saturn—the first probe to do so. In December, Europe's launching rocket Ariane makes a perfect maiden flight from Kourou, French Guiana.	**1979** Skylab breaks up in space, with debris raining down in Western Australia – fortunately in remote areas.
1980 Voyager 1 sends incredible pictures back from Saturn in November, showing its dazzling ring system and many more moons. In December, NASA launches the most powerful communications satellite ever, Intelsat V. This is capable of handling 12,000 simultaneous telephone conversations.	**1981** On April 12, the first space shuttle flight takes place. During the mission, the orbiter Columbia remains in space for a little over two days, returning to make a perfect landing at the Edwards Air Force Base in California. Astronauts John Young and Robert Crippen are aboard. In November, Columbia becomes the first space vehicle to return to space, in another perfect mission.
1982 The USSR launches Salyut 7 into orbit in April.	**1982** The Soviet cosmonaut Svetlana Savitskaya becomes the second woman to venture into space, traveling to Salyut 7 for a week-long stay in August.
1983 The international probe IRAS is launched in January. It makes many interesting discoveries during its 10 months in space, such as observing new stars being born. In June, the U.S. probe Pioneer 10 starts its journey into interplanetary space, traveling beyond the orbit of the outermost planet.	**1983** The second shuttle orbiter, Challenger, takes to the skies in April. In June, Sally Ride becomes the first U.S. woman astronaut. In November, Spacelab makes its first flight in the orbiter Columbia. Six astronauts are on board. **1984** U.S. astronaut Bruce McCandless makes the first untethered spacewalk, "flying" the MUU in February. In April, astronauts in MUUs help capture and repair the satellite Solar Max. Also in that month, the cosmonauts Leonid Kizim, Vladimir Solovyov and Oleg Atkov return to earth after a record stay in Salyut 7 of 237 days. **1985** First flight of the fourth space shuttle orbiter Atlantis takes place in October. In November, U.S. astronauts carry out construction work in orbit, rehearsing procedures that will have to be carried out when NASA's space station is built in the 1990s.
1986 The U.S. probe Voyager 2 transmits images of Uranus over a distance of some 2,000 million miles (3,000 million km) in January. In March, the European probe Giotto sends back pictures of the heart of Halley's comet, which proves to be a potato-shaped lump of rock about 10 miles (15 km) long and about 5 miles (8 km) across.	**1986** The shuttle orbiter Challenger explodes during take-off, killing all seven crew members. Meanwhile, the USSR pushes ahead with its space station program by launching Mir, a base unit for a large space complex. **1987** In June the first purpose-built scientific module docks with Mir. It is called Kvant.

The 1980s

GLOSSARY OF TERMS

Abort Cut short a spaceflight.

Airlock An airtight cabin through which astronauts enter or leave a spacecraft.

Antenna Another term for aerial.

Apollo The American project that launched astronauts to the moon. The first moon landing took place on July 20, 1969; the last astronauts left the moon on December 14, 1972.

Ariane The main launching vehicle for the European Space Agency. It lifts off from a launch site near Kourou in French Guiana, South America.

Artificial gravity A force similar to gravity produced by rotating a space station.

Astronaut A person who journeys into space.

Atmosphere The layer of gases around the earth. such as the air around the Earth.

Bio-sensors Devices attached to an astronaut's body to sense such things as pulse rate and breathing.

Blackout A breakdown in communications between a spacecraft and ground control during re-entry. The term also means a loss of consciousness when an astronaut experiences high G-forces.

Blast-off Launching of a rocket.

Booster The first stage of a launching rocket.

Burn The period when a rocket is firing.

CapCom Short for Capsule Communicator—the person who talks to astronauts during spaceflight.

Capsule The name given to the cramped crew cabin of early spacecraft.

Centrifuge A machine that whirls an astronaut around and around during training. It produces the kind of forces the astronaut will experience at lift-off and reentry.

Cosmic rays Energetic particles that bombard the earth from outer space.

Cosmonaut Russian word for astronaut, or space traveler.

Countdown The period of time before a rocket launch during which a certain sequence of events must be carried out.

Docking The joining together of one spacecraft with another.

Below: Apollo astronaut on the Moon, with lunar module and lunar rover.

Above: Ariane on the launch pad.

Drag The resistance a spacecraft experiences when traveling through the air during reentry.

ESA The European Space Agency —the body responsible for space activities in Europe.

Escape velocity The speed a body must have in order to escape from the earth's gravity. It is 25,000 mph an hour (40,000 km an hour).

EVA Short for extra-vehicular activity, which is the correct name for space-walking.

Extraterrestrial Not living on or belonging to the Earth.

Free fall The condition astronauts experience in orbit, when they and everything else are "falling around the earth" and there appears to be no gravity. We commonly call this state "weightlessness."

Fuel cell A special battery used in manned spacecraft that produces electricity by chemically burning hydrogen and oxygen.

G-forces The forces an astronaut experiences during launching and reentry, when the rocket changes speed suddenly.

Gravitation A force that causes one body to pull, or attract, another. The more massive a body, the greater its power of attraction.

Gravity The pull of the earth on any nearby body.

An American Viking probe took this picture of Mars in 1976.

Heat shield A layer on the outside of a spacecraft that protects it from heat. Heat is produced by air friction when a spacecraft reenters the earth's atmosphere. The shuttle has a thick layer of tiles to protect it.

Hold A temporary halt in the countdown before a launching.

Interplanetary Between planets.

Interstellar Between the stars.

Launch pad The place from which a rocket takes off.

Life-support systems The system in a spacecraft that keeps its crew alive and well.

Lift-off The moment when a rocket lifts itself off the launch pad.

Lunar Relating to the moon.

Meteorite A piece of rock from outer space that falls to the earth.

Meteoroids Particles of dust and rock that travel in space.

MMU Short for Manned Maneuvering Unit, the shuttle astronaut's jet-propelled backpack.

Mock-up A dummy spacecraft that looks like the real thing.

Module A section of a spacecraft; or a package of instruments.

NASA The main American space authority. NASA stands for National Aeronautics and Space Administration.

Orbit The curved path of a body in space around another body.

Orbital velocity The speed a body must maintain to remain in orbit. The orbital velocity 150 miles (250 km) above the earth is about 17,500 mph (28,000 km an hour).

Oxidizer The propellant in a rocket that provides the oxygen to burn the fuel.

Payload The "cargo" a space vehicle carries.

Planet A heavenly body that circles around a star. The earth and

Launch pad for the American shuttle at the Kennedy Space Center, Florida.

Weightless astronauts pose for a topsy-turvy picture in Spacelab.

Mars are planets that circle the Sun.

Probe An unmanned spacecraft sent to investigate distant bodies in space, such as the planets.

Propellants The substances burned by rockets to produce the gases which propel them forward. Generally there are two propellants, fuel and oxidizer.

Radiation Rays and particles given out by the sun and the stars, and also by certain rocks and metals (such as uranium).

Re-entry The moment when a spacecraft reenters the earth's atmosphere after a space flight.

Retro-rocket A rocket that is fired forward to slow down a spacecraft.

Rollout Moving a launching vehicle out to the launch pad.

Rocket An engine propelled forward by a stream of gases shooting out backward. Unlike a jet engine, a rocket carries its own oxygen and so it can work in space.

Satellite A body that circles around a planet. The earth has one natural satellite, the moon, and many artificial satellites.

Simulator A full-scale model of (for example) a spacecraft, which looks and acts like the real thing. Used in training astronauts.

Solar Relating to the sun.

Solar cell A special battery used to power spacecraft that changes the energy in sunlight into electricity.

Space sickness The nausea astronauts often experience when they go into space. It usually wears off after a few days.

Spacewalk The common term for what is properly called extravehicular activity, or activity outside a spacecraft.

Sputnik The Russian word for satellite. In particular it refers to Sputnik 1, the first artificial satellite, which the USSR launched into orbit on October 4, 1957 to begin the Space Age.

Stage Part of a step-rocket that fires and then falls away.

Stellar Relating to the stars.

Step-rocket An arrangement of three or more rockets on top of one another. Each rocket stage fires and then falls away in turn.

Telemetry Measuring from a distance. On satellites, things are measured by instruments and then the results are radioed back to earth.

Terrestrial Relating to the earth.

Tracking Following the path of a spacecraft as it travels through space, using radio, radar or photography.

Trajectory The path of a body through space.

Umbilical A connection carrying cables or tubes between (for example) a rocket and the launch tower. Early astronauts spacewalked in spacesuits supplied with oxygen through an umbilical.

Weightlessness A condition astronauts experience in orbit, when there is no gravity to keep their feet "on the ground."

Zero-G Another term for weightlessness.

INDEX

PHOTOGRAPHIC ACKNOWLEDGEMENTS
The publishers wish to thank the following for supplying photographs for this book: Page 2 Radio Times Hulton Picture Library; 3 PHOTRI/ZEFA; 5 PHOTRI/ZEFA *bottom left*; 7 NASA; 11 Spacecharts/NASA; 14–15 Spacecharts/NASA; 19 PHOTRI/ZEFA *top*, Spacecharts/NASA *bottom*; 22 Spacecharts/NASA; 28–30 Spacecharts/NASA.